Axolotl Care Guide for Beginners

Complete step-by-step guide on how to take care of Axolotl, including breeding, caring, feeding, and aquarium setup.

Lucy Howard

Copyright@2024 All Rights Reserved: This book may not be reproduced, transcribed, photocopied, or transmitted without permission from the publisher or the author. This publication is copyright protected and intended for personal use only with no means of plagiarizing, lifting, or reproduction to be resold as a different book. The information in this book is for educational and reference purposes only. All sincerity has been implemented to produce correct, relevant, accurate, and reliable information. Using this book allows readers to comply that no liability is held against the author for any direct or indirect costs resulting from the informational content. But no limitation to possible errors, inaccuracies, and mistakes.

Table of Contents

AXOLOTL CARE GUIDE FOR BEGINNERS I

INTRODUCTION 1

    How to Use this Book? 1

CHAPTER 1: UNDERSTANDING AXOLOTLS 3

    History 3
    Natural Habitat of Axolotls 4
    Why Have Axolotls Evolved This Way? 5
    What Do Axolotls Eat and How Long Do They Live For? 6
    Are Axolotls Endangered Species? 7

CHAPTER 2: AXOLOTL SPECIES AND COLOR VARIATIONS 9

    Axolotls Species 9
    Ambystoma mexicanum 9
    Physical Characteristics: 10
    Related Species in the Ambystoma Genus 11
    Ambystoma tigrinum (Tiger Salamander) 11
    Ambystoma mavortium (Barred Tiger Salamander) 12
    Ambystoma maculatum (Spotted Salamander) 13
    Ambystoma opacum (Marbled Salamander) 14
    Importance of Ambystoma mexicanum 15
    Color Variations of Axolotls 16
    Wild Type Axolotl 16
    Leucistic Axolotl 17
    Genetics 18
    Albino Axolotl 19

- Genetics 20
- Golden Albino Axolotl 20
- Genetics 21
- Melanoid Axolotl 22
- Genetics 23
- Copper Axolotl 23
- Genetics 24
- GFP (Green Fluorescent Protein) Axolotl 25
- Genetics 26
- Chimera Axolotl 26
- Genetics 27
- Three main types of pigment cells influence axolotl coloration: 28

CHAPTER 3: HOW TO SET UP AN AXOLOTL TANK 29

- Tank Selection 29
- Substrate 30
- Get the right Filters 31
- Add Water to the Tank 32
- Cycle the water before placing your Axolotls inside 33
- Test the water quality 34
- Decor and Hiding Places 35
- Lighting 35
- Maintenance and Cleaning 36

CHAPTER 4: INTRODUCING YOUR AXOLOTL TO THEIR NEW HOME 37

- Final Check Before Adding Your Axolotl to the Tank 37
- Do not House Axolotls with other Fish. 38
- Acclimating Your Axolotl 38

- 1. Float the Bag: 39
- 2. Gradual Mixing: 39
- 3. Release the Axolotl: 39
- Monitoring Your Axolotl 40
- Maintaining a Healthy Environment 41

CHAPTER 5: HOW TO FEED YOUR AXOLOTL 43

- Understanding Axolotl Diet 43
- Types of Food 43
- Feeding Frequency 45
- Tips for Feeding 45
- Troubleshooting Feeding Issues 46
- Nutritional Requirements 46

CHAPTER 6: DAILY CARE, MAINTENANCE, BEHAVIOR AND INTERACTION 49

- Daily Observation and Health Check 49
- Water Quality and Tank Maintenance 50
- Cleaning and Tank Maintenance 50
- Understanding Axolotl Behavior 51
- Social Interaction 52
- Interaction with Humans 52
- Creating a Stimulating Environment 53

CHAPTER 7: HOW TO BREEDING AXOLOTLS 55

- Understanding the Breeding Process 55
- Axolotls Life Cycle 56
- Setting Up a Breeding Tank 57
- Caring for Eggs and Larvae 58
- Raising Healthy Juveniles 59

CHAPTER 8: COMMON MISTAKES TO AVOID AND FREQUENTLY ASKED QUESTIONS 61

    FREQUENTLY ASKED QUESTIONS 63

CHAPTER 9: AXOLOTLS FOR BUSINESS 69

    MARKET RESEARCH AND BUSINESS PLANNING 69

    SETTING UP A BREEDING FACILITY 70

    RAISING AND CARING FOR AXOLOTL LARVAE 70

    MARKETING AND SALES 71

    PRICING AND SALES CHANNELS 71

    CUSTOMER CARE AND SUPPORT 72

    ETHICAL BREEDING AND BUSINESS PRACTICES 73

CHAPTER 10: FUN FACTS ABOUT AXOLOTLS 74

CONCLUSION 77

    COMMON MISTAKES AND TROUBLESHOOTING 78

    ENCOURAGEMENT AND FINAL TIPS 78

    WHERE TO FIND MORE INFORMATION 78

Introduction

Axolotls are super cute fishes that need special care to live long, healthy, and happy. They've become a favorite for aquarium owners, first-time and beginners considering breeding for fun and business purposes. It is famous for aquariums because of its unique looks, relatively low-maintenance care requirements, and entertaining behaviors. Though Axolotl could be challenging for new breeders, like starting any new idea or caring for livestock, mistakes and not giving proper care are familiar to new hobbyists. This guide is written to meet the needs of first-time Axolotl owners. It offers detailed, easy-to-follow instructions with photos and insights to answer questions and avoid mistakes. The goal is to ensure you feel confident and enjoy caring for your Axolotls while they develop and bring you joy.

How to Use this Book?

This book covers the basics and intermediary knowledge of Axolotls, which is what everyone must know before considering breeding Axolotls for fun or business purposes. Then, you will find other friendly chapters, starting with history, natural habitat, Axolotls species, and color variations, breeding, setting up a breeding tank, caring for eggs, and raising juveniles. You'll also find tips on starting and managing a small axolotl breeding business, including legal considerations and marketing strategies. We hope you find this book worthwhile, engaging, and an essential companion in your journey with these incredible creatures.

Chapter 1: Understanding Axolotls

History

The Axolotls, also known as the Mexican walking fish, are significant in natural history and human culture. Native to the ancient lake complex of Xochimilco in Mexico, Axolotls have entertained people for centuries. The name "Axolotls" comes from the Nahuatl language, spoken by the Aztecs, who revered these unique creatures. In Aztec mythology, the axolotl was associated with the god Xolotl, who, according to legend, transformed into an Axolotl to avoid being sacrificed. The Axolotl's ability to regenerate lost limbs and other body parts contributed to its mythical status.

European scientists first encountered axolotls in the 19th century. The species was brought to international attention when specimens were shipped to Paris for study. Scientists were intrigued by the Axolotl's neoteny—retaining juvenile characteristics into adulthood—and its remarkable regenerative capabilities. Axolotls quickly became popular subjects in scientific research, particularly in developmental biology and regenerative medicine. Over the years, the natural habitat of Axolotls has faced severe threats due to urbanization, pollution, and the introduction of invasive species. As a result, Axolotl populations in the wild have drastically declined, leading to their classification as critically endangered. Conservation efforts are now focused on habitat restoration and captive breeding programs to ensure the survival of this extraordinary species.

Natural Habitat of Axolotls

Axolotls are native to the Xochimilco Lake complex near Mexico City, characterized by calm, clear, and nutrient-rich waters. These lakes, formed by volcanic activity, offer the ideal environment for Axolotls, with temperatures ranging from 14 to 20 degrees Celsius (57 to 68 degrees Fahrenheit). The shallow waters, dense aquatic vegetation, and clear conditions support their unique needs, providing shelter, breeding grounds, and optimal living conditions.

However, the natural habitat of Axolotls has been severely impacted by urbanization, pollution, and invasive species like tilapia and carp. These threats have drastically reduced their habitat, contributing to their critically endangered status.

The natural habitat of Axolotl

Why Have Axolotls Evolved This Way?

Axolotls exhibit several unique evolutionary traits that distinguish them from other amphibians. One of the most remarkable is neoteny, where they retain their larval features throughout their lives, remaining aquatic and keeping their gills. This adaptation is believed to be advantageous in the stable, calm, and oxygen-rich environment of the Xochimilco Lake complex, allowing them to avoid the risks of terrestrial life.

Another extraordinary trait is their regenerative abilities, enabling axolotls to regrow limbs, spinal cord parts, and even portions of their brain. This capability likely evolved as a defence mechanism to enhance their survival and injury recovery. Axolotls' specific adaptations to their natural habitat, such as their efficient gills for extracting oxygen, camouflage against the lakebed, and feeding strategies, have all contributed to their evolutionary success in Xochimilco. These adaptations highlight the importance of conserving their natural environment to ensure survival.

Evolution of Axolotls to the adult stage

What Do Axolotls Eat and How Long Do They Live For?

Axolotls are carnivorous and have a varied diet in both the wild and captivity. They feed on small fish, worms, insects, and crustaceans in their natural habitat. In captivity, they flourish on a diet of high-quality pellets, frozen or live bloodworms, brine shrimp, and earthworms. It's essential to provide a balanced diet to ensure their health and longevity. Axolotls have a relatively long lifespan for amphibians. In captivity, they can live for 10 to 15 years with proper care, while their lifespan in the wild is typically shorter due to environmental challenges and predation. Critical factors ensuring a long, healthy life for axolotls include maintaining clean, cool water, providing a suitable diet, and creating a stress-free environment.

Are Axolotls Endangered Species?

Yes, Axolotls are critically endangered. According to the IUCN Red List of Threatened Species, their population is alarmingly low, with only 50-1,000 adults remaining in the wild. The primary reasons for their endangered status are the severe diminishment of their natural habitat and the multiple threats they face within their environment. Despite the critical situation, Axolotls have not gone extinct, although it was anticipated that they might be by 2020. Conservation efforts are ongoing to protect and restore their natural habitat. These efforts include cleaning the waterways, reintroducing native plants, and establishing protected areas to support the remaining wild populations. Additionally, breeding programs in captivity aim to ensure the survival of Axolotls and potentially reintroduce them to their native habitats. While Axolotls are critically endangered, concerted conservation actions provide hope for survival.

Understanding axolotls and their needs is the first step toward providing them with the best care possible. These extraordinary creatures have much to offer, from attractive looks to marvelous regenerative abilities. As you board this journey with your Axolotl, you'll learn interesting wonders that make them one-of-a-kind pets.

Chapter 2: Axolotl Species and Color Variations

Species and color type are factors for home aquarium or breeding process, starting from the classic wild type to the rare and beautiful albino. Each color variation has its own unique and care considerations. This chapter covers the different species of Axolotls, although most pet axolotls belong to a single species. Then, the various color morphs make axolotls so visually appealing.

Axolotls Species

When discussing Axolotls, we typically refer to a single species, Ambystoma mexicanum. However, Axolotls belong to a broader group of salamanders within the Ambystoma genus, known as mole salamanders. Here, we'll examine Ambystoma mexicanum, the species commonly kept as pets and studied, and briefly touch upon other related species within the genus.

Ambystoma mexicanum

Common Name: Axolotl or Mexican Walking Fish

Natural Habitat: Lake complex of Xochimilco near Mexico City

Distinctive Traits: Axolotls are neotenic, meaning they retain their larval features, such as gills, throughout their adult life. This makes them fully aquatic, unlike many other salamanders that undergo metamorphosis and live on land as adults.

Conservation Status: Critically endangered in the wild due to habitat loss, pollution, and introduction of invasive species.

Ambystoma mexicanum

Physical Characteristics:

Size: Typically grows to about 9-12 inches (23-30 cm) in length.

Lifespan: Can live 10-15 years in captivity with proper care.

Color Morphs: Exhibit a variety of color morphs, including wild type, leucistic, albino, golden albino, melanoid, copper, and GFP.

Related Species in the Ambystoma Genus

While Ambystoma mexicanum is the most well-known due to its unique traits and popularity in the pet trade, several other species exist within Ambystoma. These species are collectively known as mole salamanders and share some characteristics but differ in life cycles and habitats.

Ambystoma tigrinum (Tiger Salamander)

Habitat: Widely distributed across North America.

Life Cycle: Unlike Axolotls, tiger salamanders undergo metamorphosis and live part of their lives on land.

Physical Traits: Known for their distinctive tiger-like striping and robust build.

Conservation Status: Least concern, although local populations can be threatened by habitat loss.

Ambystoma tigrinum (Tiger Salamander)

Ambystoma mavortium (Barred Tiger Salamander)

Habitat: Found in the western United States.

Life Cycle: Like the tiger salamander, it undergoes metamorphosis and is aquatic and terrestrial.

Physical Traits: Characterized by their barred patterns and similar robust build to other tiger salamanders.

Conservation Status: Least concern, though also subject to habitat pressures.

Ambystoma mavortium (Barred Tiger Salamander)

Ambystoma maculatum (Spotted Salamander)

Habitat: Found in forests throughout the eastern United States and Canada.

Life Cycle: Metamorphoses into a terrestrial adult but returns to water to breed.

Physical Traits: Noted for their striking yellow or orange spots on a dark background.

Conservation Status: Least concern, with stable populations in appropriate habitats.

Ambystoma maculatum (Spotted Salamander)

Ambystoma opacum (Marbled Salamander)

Habitat: Inhabits moist woodlands in the eastern United States.

Life Cycle: Lays eggs in dry vernal pools that hatch when rains come, with adults being terrestrial.

Physical Traits: Features bold black and white marbling patterns.

Conservation Status: Least concern, though sensitive to environmental changes.

Ambystoma opacum (Marbled Salamander)

Importance of Ambystoma mexicanum

The axolotl (Ambystoma mexicanum) stands out among its relatives due to its unique evolutionary traits and adaptability to captivity, making it a subject of attraction for hobbyists and scientists. Its ability to remain in its larval stage and regenerate body parts has made it a model organism for studying development, regeneration, and potential medical applications.

Color Variations of Axolotls

Understanding the different color variations of axolotls can enhance your appreciation of these fascinating creatures and help you select the perfect axolotl for your aquarium or breeding project. The color variations range from the naturally camouflaged wild type to the strikingly beautiful leucistic and albino morphs, each with its own unique genetic background and care considerations.

In this section, we'll examine the most popular and widely recognized color morphs of Axolotls, exploring their physical characteristics, genetics, and the unique beauty each variation brings to the world of amphibians. Whether you're a hobbyist looking to expand your collection or a first-time axolotl owner, understanding these color variations will provide valuable insights into the captivating world of axolotls.

Here are some of the most popular and widely recognized color morphs:

Wild Type Axolotl

Wild Type Axolotls, belonging to Ambystoma mexicanum, are characterized by their dark, mottled appearance with shades of green, brown, and black.

This natural coloration, is due to a mix of melanophores, xanthophores, and iridophores, which helps them blend into their native habitat. In the wild, wild-type Axolotls flourish in cool, clear waters with dense vegetation, feeding on various aquatic organisms. They are nocturnal solitary and use their camouflage to ambush prey. Breeding involves males depositing spermatophores for females to fertilize their eggs, which are laid on submerged vegetation.

An image of a Wild type Axolotl

Leucistic Axolotl

Leucistic axolotls have a pale pink or white body. Their skin's lack of dark pigmentation gives them a clean and delicate appearance, often mistaken for albinos. Their gills are usually bright red or pink, standing out vividly against their light bodies. The color of the gills can vary depending on blood flow and health. One of the defining features of leucistic axolotls is their dark, black eyes. This is a significant distinction from albino axolotls with red or pink eyes. Like other morphs, Leucistic typically grow to about 9-12 inches (23-30 cm) in length.

An image showing Leucistic Axolotl

Genetics

The leucistic coloration is caused by a genetic mutation that reduces all types of pigment cells. This mutation affects melanophores (black/brown), xanthophores (yellow), and iridophores (reflective), leading to their pale appearance. Unlike albinism, leucism does not completely eliminate pigment in the eyes, so leucistic Axolotls retain their dark, black eyes.

Albino Axolotl

Albino axolotls are easily distinguished by their lack of pigmentation, giving them a striking appearance. Their bodies are usually pale, translucent, white, or yellow and often appear pink due to the visibility of their blood vessels and internal organs through their skin. This translucent quality gives them a delicate and unique look. Their gills are typically bright red or pink, standing out vividly against their light bodies. The color of the gills can vary depending on blood flow and health, adding to their striking appearance.

One of the defining features of albino axolotls is their red or pink eyes. This is a significant distinction from leucistic axolotls with dark, black eyes. The lack of melanin in albino axolotls gives them these unique eye colors, contributing to their distinct and ethereal appearance. Like other morphs, albino axolotls typically grow to about 9-12 inches (23-30 cm) in length.

An image showing Albino Axolotl

Genetics

The albino coloration is caused by a genetic mutation that eliminates melanin production in the skin, eyes, and other tissues. This mutation affects melanophores (black/brown pigment cells), making them translucent. Unlike leucism, which reduces all pigment cells, albinism eliminates melanin, resulting in the characteristic pink or red eyes and the pale, translucent body of albino Axolotls.

Golden Albino Axolotl

Golden albino Axolotls are attractive due to their unique,

shimmering golden-yellow bodies. Their skin ranges from a pale yellow to a rich, metallic gold, giving them a radiant and eye-catching appearance. Their gills are bright red or pink, providing a vivid contrast to their golden bodies. This striking feature varies with blood flow and health. One defining characteristic of golden albino Axolotls is their red or pink eyes, setting them apart from leucistic axolotls, which have dark eyes. Golden albino Axolotls typically grow to about 9-12 inches (23-30 cm) in length.

An image showing Golden Albino Axolotl

Genetics

The golden coloration in albino Axolotls is due to a genetic mutation that eliminates melanin production.

This mutation affects melanophores (black/brown pigment cells), giving them a golden appearance and pink or red eyes.

Melanoid Axolotl

Melanoid Axolotls are unique due to their dark, uniform coloration. Their bodies are typically a deep, solid black or dark brown, lacking the iridescence seen in other Axolotl morphs. This gives them a sleek and striking appearance that is quite distinctive. Their gills are also usually dark, often appearing black or deep red. The uniformity in their pigmentation extends to their gills, enhancing their overall dark appearance. One defining characteristic of melanoid Axolotls is their dark eyes, which lack the reflective iridophores in other morphs. Like other morphs, melanoid Axolotls typically grow to about 9-12 inches (23-30 cm) in length.

An image showing Melanoid Axolotl

Genetics

Melanoid coloration is caused by a genetic mutation affecting pigmentation in the skin and eyes. This mutation eliminates iridophores (reflective pigment cells) and reduces the number of xanthophores (yellow pigment cells), resulting in a uniform dark coloration.

Copper Axolotl

Copper Axolotls are distinguished by their warm, earthy tones. Their bodies exhibit light to dark brown shades, often with a coppery, metallic sheen that gives them a unique and appealing appearance. This coloration sets them apart from the more common morphs and adds a touch of warmth to their look. Their gills are usually a darker shade of brown or reddish-brown, providing a subtle yet attractive contrast to their body color. The gill color can vary depending on their health and blood flow. One defining feature of copper axolotls is their light-colored eyes, ranging from golden to light brown. Like other morphs, copper Axolotls typically grow to about 9-12 inches (23-30 cm) in length.

An image showing Copper Axolotl

Genetics

The copper coloration is due to a genetic mutation that affects the pigmentation in their skin and eyes. This mutation alters the distribution and type of pigment cells, resulting in the warm, brown tones characteristic of copper Axolotls.

GFP (Green Fluorescent Protein) Axolotl

GFP axolotls are remarkable for their ability to glow green under ultraviolet (UV) light, thanks to the presence of the green fluorescent protein gene. They may appear like any standard morph in normal lighting, but they exhibit a vibrant green fluorescence under UV light, making them an interesting and visually stunning variety. Like the rest of their bodies, their gills glow green under UV light. This feature adds to the overall luminous effect, creating a striking and captivating display. Growth is typically about 9-12 inches (23-30 cm) in length.

An image showing (Green Fluorescence Proptein) Axolotl

Genetics

The GFP coloration is due to the insertion of the green fluorescent protein gene, originally derived from jellyfish, into their genome.

This genetic modification allows the Axolotl's cells to produce the fluorescent protein, which glows green when exposed to UV light.

Chimera Axolotl

Chimera Axolotls are one of the most visually striking and rarest morphs, known for their unique and unusual appearance. These Axolotls have distinct, split-color patterns resulting from merging two different genetic lines. Typically, their bodies are divided into two different colors, often split right down the middle, creating a dramatic and eye-catching contrast. Their gills also reflect this split coloration, often mirroring the body's two-toned pattern. This feature adds to their extraordinary appearance, making each chimera axolotl unique. chimera axolotls typically grow to about 9-12 inches (23-30 cm) in length.

An image showing Chimera Axolotl

Genetics

The chimera coloration is the result of the fusion of two different embryos at an early stage of development. This leads to an organism that contains cells from both genetic lines. The phenomenon is rare and occurs naturally, leading to a distinct and dramatic split in coloration. The resulting axolotl has tissues that express two different sets of genetic information, creating its characteristic split appearance. The diverse color morphs of Axolotls are an attractive result of their genetics.

Three main types of pigment cells influence axolotl coloration:

Melanophores: Contain black or brown pigment (melanin).

Xanthophores: Contain yellow pigment.

Iridophores: Contain reflective or iridescent pigment.

The interaction and presence of these pigment cells and genetic mutations result in various color morphs. For example, the absence of melanophores results in albino Axolotls, while the lack of iridophores and a high concentration of melanophores create melanoid Axolotls. Selective breeding in captivity has allowed enthusiasts to enhance and propagate these color traits, leading to the wide variety of beautiful and unique axolotls we see today.

Understanding the genetics behind these traits can help breeders produce specific color morphs and ensure healthy genetic diversity in their populations.

Chapter 3: How to Set Up an Axolotl Tank

Axolotls have specific needs regarding their habitat, and understanding these requirements will help you set up an ideal tank. Creating the perfect environment for your Axolotl is essential for their health, well-being, and longevity. This chapter will guide you through selecting the right tank, choosing the appropriate equipment, and maintaining optimal water conditions to ensure your Axolotl flourishes.

Tank Selection

For a single Axolotl that is 1-5 inches long, a minimum tank size of 20 gallons is. If you plan to keep more than one Axolotl, increase the tank size by at least 10 gallons per additional Axolotl. A larger tank provides more swimming space and helps maintain water quality by diluting waste products. Axolotls prefer a tank with a large surface area and ample floor space. Long, shallow tanks are better than tall, narrow ones, allowing for more horizontal swimming.

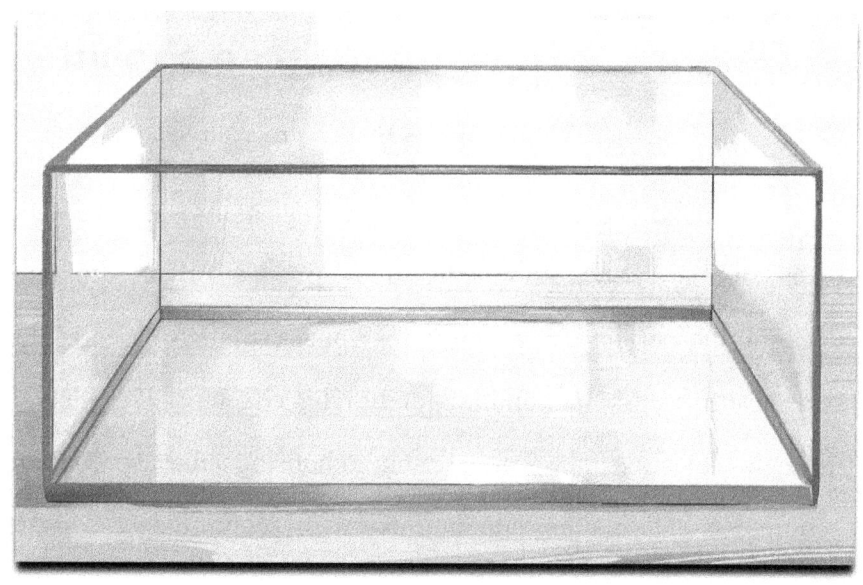

Ideal Axolotl tank size

Substrate

Your Axolotl will need sand to lay at the tank's base and decorations, such as rocks or caves. A bare-bottom tank is easy to clean and prevents the risk of impaction from swallowing substrate. However, it can look less natural and may not provide enough traction for Axolotls. Fine sand is a safe and popular choice for Axolotl tanks. It mimics their natural environment and reduces the risk of impaction. Ensure the sand is free of sharp particles that could harm your Axolotl. Avoid using gravel, as Axolotls can accidentally ingest it while feeding, leading to serious health issues.

Image of well-set-up tank

Get the right Filters

Sponge Filters: Gentle on Axolotls and provide both mechanical and biological filtration. Ideal for smaller tanks and young axolotls.

Canister Filters: Offer excellent filtration for larger tanks and can handle higher bio-loads. Ensure the outflow is gentle to avoid stressing the Axolotl. Hang-on-Back (HOB) Filters: Suitable for medium to large tanks, but ensure the flow is adjustable to prevent strong currents. A sponge filter is best because it filters particles, helps beneficial bacteria grow, and provides oxygen and water circulation to pick up loose debris.

Filtration Tips:

Aim for a filter that can process the entire tank volume 4-6 times per hour. Use filter media that promotes beneficial bacteria growth, such as sponges, ceramic rings, or bio balls. Regularly clean the filter media to maintain optimal filtration efficiency without disrupting the beneficial bacteria colony.

Add Water to the Tank

A water conditioner is essential because tap water contains chlorine, which is toxic to Axolotls and other freshwater aquatic animals. The conditioner (a chemical solution that neutralises harmful substances in tap water) removes chlorine through chemical reactions. Be sure to follow the instructions on the product label, as the amount needed can vary. Axolotls flourish in cooler water temperatures, ideally between 60-68°F (16-20°C). Avoid temperatures above 72°F (22°C), which can cause stress and health issues. Cycle the water before placing your Axolotls inside. Cycling the water converts ammonia to nitrite, producing beneficial bacteria that maintain a healthy tank environment. To cycle a tank, add a source of ammonia, such as household ammonia, and let the tank run for 6 to 8 weeks. This allows beneficial bacteria to build up in the filter and substrate. Be sure to prepare in advance. The tank is ready once the water conditions stabilize. The pH should be between 6.5 and 8, with 7.4 to 7.6 ideal.

An example of a conditioned water

Cycle the water before placing your Axolotls inside

Before introducing your Axolotls to the tank, it's crucial to cycle the water. This process allows ammonia to convert into nitrite and then nitrate, fostering beneficial bacteria that help maintain a healthy environment. To cycle a tank, add a small amount of ammonia, such as household ammonia, and let the tank run for about 6 to 8 weeks. During this time, beneficial bacteria will develop in the filter and substrate. You'll know the tank is fully cycled when the water parameters stabilize. The ideal pH for an Axolotl tank is between 6.5 and 8, with 7. 4 to 7.6 being

Tank water cycle in progress

Test the water quality

Even if the water looks clean, it might still be unsuitable for your axolotls. Use a water tester kit to check the water quality weekly. Compare the results to suitable charts to ensure the water is safe for your Axolotls.

Water quality test

Decor and Hiding Places

Provide multiple hiding spots using caves, PVC pipes, or aquarium-safe decorations. Axolotls need places to retreat and feel secure. Live or artificial plants can enhance the tank's appearance and provide additional hiding spots. Choose plants that tolerate cooler water temperatures, such as java fern, anubias, or moss balls.

Lighting

Axolotls do not require strong lighting and prefer dimly lit environments. Too much light can stress them and promote algae growth. Use low-intensity aquarium lights, and consider using a timer to mimic natural day and night cycles. Ensure there are shaded areas in the tank where the Axolotl can retreat from the light.

Maintenance and Cleaning

Perform 20-30% weekly water changes to maintain water quality and remove waste products. Regularly clean the tank walls, substrate, and decorations to prevent algae buildup and maintain a clean environment. Clean filter media regularly, but avoid washing it with tap water, as this can kill beneficial bacteria. Rinse it in tank water during water changes instead.

Following these guidelines and maintaining a clean, well-equipped tank can create an ideal environment for your Axolotl to flourish. A properly set up tank ensures your pet's health and happiness and enhances your enjoyment as you observe their unique and attractive behaviors.

Chapter 4: Introducing Your Axolotl to Their New Home

Introducing your axolotl to its new home is exciting, but it requires careful preparation and patience to ensure a smooth transition. This chapter will guide you through the steps to acclimate your Axolotl to the tank environment, minimize stress, and promote a healthy start. From temperature matching to gradual water introduction, you'll learn the best practices to help your Axolotl settle comfortably into its new habitat.

Final Check Before Adding Your Axolotl to the Tank.

Ensure the tank has been properly cycled for 6-8 weeks with stable water parameters. Test the water quality one more time using a water tester kit. Check for ammonia, nitrite, nitrate, and pH levels. The ideal pH for axolotls is between 7.4 and 7.6. Confirm that the water temperature is within the safe range of 60-64°F (16-18°C). Ensure the filter, heater (if needed), and aeration devices function correctly. Arrange the decorations, hides, and plants in a way that provides ample hiding spots and swimming space for your Axolotl. Double-check that there are no sharp edges or hazardous items in the tank.

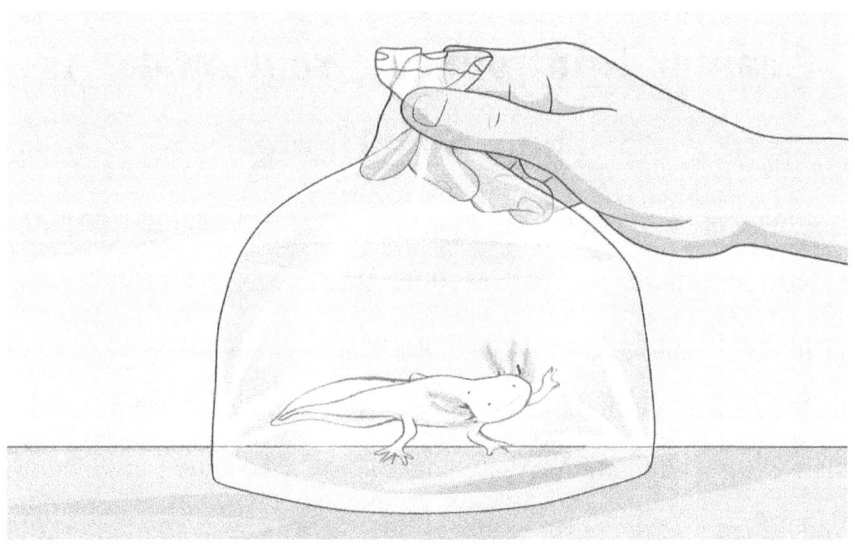

Baby Axolotl in a bag

Do not House Axolotls with other Fish.

Axolotls don't get along well with other fish because of their poor vision; they'll bite at any movement and may eat the fish. For the safety of both the fish and the Axolotl, it's best to keep only Axolotls of the same size in the tank.

Acclimating Your Axolotl

Acclimating your Axolotl to its new tank is critical to minimize stress and avoid shock from sudden changes in water conditions. Follow these steps:

1. Float the Bag:

When you bring your Axolotl home, keep it in the bag it came in. Float the bag in the tank for 15-20 minutes. This allows the temperature in the bag to gradually match the tank water temperature.

2. Gradual Mixing:

After 15-20 minutes, slowly open the bag and add some tank water. Repeat this process every 5 minutes for about 30 minutes. This helps your Axolotl adjust to the tank's water chemistry.

3. Release the Axolotl:

Gently release your Axolotl into the tank. Avoid pouring the water from the bag into the tank, as it may contain contaminants. Use a net to transfer your Axolotl if needed.

Baby Axolotl is undergoing acclimating

Monitoring Your Axolotl

After introducing your Axolotl to the tank, closely monitoring its behavior and health is important. Observe your Axolotl for signs of stress, such as excessive gill movement, lethargy, or erratic swimming. Check the water parameters and ensure an optimal environment if you notice any of these signs. Wait for a few hours before offering food to your Axolotl, as it may need time to settle in. Start with small portions and observe how well it eats, gradually increasing the amount of food as it becomes more comfortable. Continue to test the water weekly to maintain optimal conditions.

Regularly inspect your Axolotl for signs of illness or injury to ensure health and well-being.

Maintaining a Healthy Environment

Maintaining a healthy environment for your Axolotl involves regular water testing, routine maintenance, and careful observation. Regularly test the water quality to ensure that ammonia, nitrite, nitrate, pH, and temperature levels remain stable. Weekly water changes of about 20-30% are performed to keep the water clean, and periodically clean the substrate, decorations, and filter media without disturbing the beneficial bacteria. Keep a close eye on your Axolotl's health and behavior, watching for signs of illness such as changes in appetite, discoloration, or unusual swimming patterns. Early detection and prompt action can prevent serious health issues.

Introducing your Axolotl to its new tank is a crucial step in ensuring its health and happiness. Carefully preparing the tank, acclimating your Axolotl gradually, and maintaining a stable environment, you set the foundation for a succeeding aquatic pet. Remember to regularly monitor your Axolotl's behavior and water conditions to catch any issues early. Your Axolotl will adapt to its new home and flourish with patience and attention to detail. Your dedication during this initial phase will pave the way for a rewarding experience with your unique and enthralling companion.

Chapter 5: How to Feed Your Axolotl

Feeding your Axolotl properly is essential for its health and well-being. Understanding their dietary needs and feeding habits will help them receive the right nutrients and remain happy and healthy. This chapter will cover the types of food suitable for Axolotls, how often to feed them, and tips for maintaining a balanced diet.

Understanding Axolotl Diet

Axolotls are carnivorous and have specific dietary needs. In the wild, they feed on small aquatic creatures like worms, insects, and small fish. Providing a varied diet that mimics their natural feeding habits in captivity is important for their overall health.

Types of Food

Live Foods:

Earthworms: One of the best staple foods for axolotls due to their high nutritional value. Cut them into smaller pieces for younger Axolotls.

Bloodworms: Ideal for younger Axolotls or as an occasional treat for adults. These can be purchased frozen or live.

Daphnia and Brine Shrimp: Suitable for juvenile Axolotls, providing essential nutrients for growth.

Frozen Foods:

Frozen Bloodworms and Brine Shrimp: Convenient and nutritious options. Thaw before feeding to prevent digestive issues.

Frozen Beef Heart and Fish Fillets: These can be offered occasionally as a treat. Ensure they are free from preservatives and additives.

Pellets:

Axolotl Pellets: Specially formulated for Axolotls, providing a balanced diet. Choose high-quality pellets that sink to the bottom of the tank.

Other Foods:

Insects: Small, gut-loaded crickets or mealworms can be offered occasionally. Remove any uneaten insects to prevent water contamination.

Small Fish: Feeder fish can be offered occasionally, but be cautious of parasites. Only use feeder fish from a trusted source.

Feeding Frequency

Juvenile Axolotls: Feed daily to support rapid growth and development. Offer small portions of food they can consume in a few minutes.

Adult Axolotls: Feed 2-3 times per week. Adjust the quantity based on their size and appetite. Monitor their weight to ensure they are not underfed or overfed.

Tips for Feeding

Provide a diverse diet to ensure your Axolotl receives all necessary nutrients. Rotate between different types of food to prevent dietary deficiencies. Watch how your axolotl eats and adjust portions accordingly. Remove any uneaten food after 10-15 minutes to maintain water quality. Use feeding tongs or tweezers to offer food directly to your Axolotl. This method helps monitor their intake and prevents food from scattering in the tank.

Overfeeding can lead to obesity and water quality issues. Stick to a regular feeding schedule and monitor your Axolotl's body condition.

Occasionally, you can dust live food with calcium or vitamin supplements to boost their nutritional intake, especially for breeding or growing axolotls.

Troubleshooting Feeding Issues

If your Axolotl refuses to eat, check the water quality and temperature. Stress or illness can also cause a loss of appetite. Consult a veterinarian if the issue persists. Axolotls can be messy eaters, scattering food around the tank. Consider feeding in a separate container or using a feeding dish to minimize waste. Some axolotls may be sensitive to certain foods. If you notice any adverse reactions, try different food types and consult with an expert if necessary.

Nutritional Requirements

Protein: Axolotls require a high-protein diet to support growth and development.

Fat: A moderate amount of fat is necessary for energy.

Vitamins and Minerals: Ensure their diet includes essential vitamins and minerals for overall health. Feeding your Axolotl, the right diet is essential for its health and longevity.

You can ensure your Axolotl flourish by understanding their dietary needs, choosing appropriate foods, and establishing a consistent feeding routine. Regularly monitor their feeding habits and adjust as needed to keep them happy and healthy. With the right care and attention, your Axolotl will enjoy a well-balanced diet and flourish in its environment.

Chapter 6: Daily Care, Maintenance, Behavior and Interaction

Taking care of your Axolotl daily involves more than just feeding and cleaning. Regular observation, maintenance, and interaction are essential to ensure your axolotl remains healthy and happy. This chapter will guide you through the daily routines and tasks necessary to maintain an optimal environment for your axolotl, help you understand its behavior, and provide tips for safe and enjoyable interaction. Dedicating a bit of time each day to your Axolotl can build a strong bond and create a successful habitat for this intriguing creature.

Daily Observation and Health Check

Each morning, take a few minutes to observe your Axolotl. Look for changes in behavior, appearance, or activity level, as healthy axolotls are usually active, alert, and responsive. Make sure your Axolotl is eating well; uneaten food can be a sign of stress or illness. A healthy axolotl will explore its tank and interact with its environment. Additionally, observe your axolotl's gills and breathing patterns, as rapid or labored breathing can indicate poor water quality or health issues. Check for any signs of injury, discoloration, or unusual growth to ensure your Axolotl remains healthy.

Water Quality and Tank Maintenance

Regularly testing the water parameters is crucial to ensure a healthy environment for your Axolotl. Use a water test kit daily to check ammonia, nitrite, nitrate, pH, and temperature levels. Both ammonia and nitrite should be at 0 ppm, nitrate should be below 20 ppm, and pH should range between 6.5 and 8, with an ideal range of 7.4 to 7.6. The temperature should remain between 60-64°F (15-18°C). Perform a partial water change (10-20%) weekly to maintain water quality by removing waste and replenishing essential minerals. Always use a water conditioner to dechlorinate tap water before adding it to the tank.

Cleaning and Tank Maintenance

Perform daily and weekly maintenance tasks to keep your Axolotl tank clean and healthy. Daily cleaning involves using a turkey baster or siphon to remove any visible waste or uneaten food from the substrate. Weekly clean decorations and plants to eliminate algae and debris. For filter maintenance, rinse the filter media in tank water (not tap water) during water changes to preserve beneficial bacteria and ensure the filter runs smoothly,

providing adequate water circulation and oxygenation. Check the thermometer daily to ensure the tank's temperature remains stable. If you use a heater or chiller, ensure these devices function correctly to regulate the tank temperature. Maintain a consistent light cycle of 10-12 hours per day to mimic natural conditions and avoid direct sunlight to prevent temperature fluctuations. Keep a daily log of your Axolotl's feeding, behavior, and any water quality changes. This helps track patterns and identify potential issues early, ensuring your axolotl stays healthy and happy.

Understanding Axolotl Behavior

Axolotls are generally more active during the evening and nighttime, as they are nocturnal creatures that prefer to explore and hunt for food when the lights are dim. During the day, they might rest more and stay in shaded areas or hides. These curious animals enjoy exploring their environment, often moving around the tank, investigating decorations, or burrowing in the substrate. Providing a variety of hides, plants, and other tank accessories can enrich their environment and stimulate their natural behaviors.

Social Interaction

Axolotls are solitary by nature and do not require companionship. In fact, housing multiple axolotls together can sometimes lead to stress or aggression, especially if the tank is not large enough. If you do decide to keep more than one Axolotl, make sure the tank is spacious, with plenty of hiding spots to reduce territorial disputes.

Tank Mates

Choosing appropriate tank mates for your Axolotl is essential to prevent stress and injury. Small, non-aggressive fish that can tolerate similar water conditions might coexist peacefully, but avoiding adding fish to an Axolotl tank is generally best. Invertebrates like snails and shrimp can be good alternatives, although they might occasionally become a snack.

Interaction with Humans

While Axolotls can tolerate brief handling, it's best to minimize this to reduce stress. If you need to handle your Axolotl, ensure your hands are clean and wet, as dry hands can damage their delicate skin. Use a soft net or container to move them when necessary. Hand-feeding your Axolotl can be a rewarding experience; use feeding tongs or your fingers to offer food directly to your Axolotl. This can help build trust and allows you to monitor their eating habits closely. Always be gentle and patient, as sudden movements can startle them. Observing your axolotl can help you understand their unique personality and detect any signs of illness early. Watch for changes in behavior, appetite, and appearance. Keeping a daily log of their activities and health can be beneficial.

Creating a Stimulating Environment

Creating an exciting environment is essential for Axolotl's mental well-being. Periodically rearrange the tank decorations to offer new exploration opportunities and introduce safe, interactive elements like floating toys or various substrate textures. Varying how you offer food can also be enriching; try hiding food in different spots or using feeding puzzles to challenge your Axolotl. Live food can be especially engaging, allowing them to practice their hunting skills. Ensure your tank lighting mimics natural day and night cycles, avoiding bright lights since Axolotls prefer dim environments. Providing shaded areas and hiding spots will help them feel secure and reduce stress.

In conclusion, maintaining the health and happiness of your

Axolotl requires consistent daily care, regular maintenance, and a deep understanding of their behavior and interaction needs. maintaining a routine of monitoring water quality, performing necessary maintenance, and providing a stimulating environment ensures a successful habitat for your Axolotl. Observing their behavior and interacting with them thoughtfully further enriches their lives and strengthens your bond. With these practices, you can create a balanced and enjoyable environment that supports the well-being of your unique and beloved pet.

Chapter 7: How to Breeding Axolotls

Breeding Axolotls can be a rewarding yet intricate process that requires careful planning and attention to detail. This chapter covers the essentials of breeding Axolotls, from understanding their breeding behavior to setting the right conditions and caring for the offspring. Even though you are rearing Axolotls for personal enjoyment or as part of a more extensive breeding program, this guide will provide you with the knowledge needed to succeed.

Understanding the Breeding Process

Axolotls reach sexual maturity at 12-18 months. Breeding typically occurs when both male and female Axolotls are healthy and of suitable size. Males and females should be introduced to each other in a separate breeding tank to encourage mating. During the breeding season, males perform a unique courtship dance known as a "spermatophore" display, where they deposit a sperm packet on the tank floor. The female then picks up this packet with her cloaca to fertilize her eggs internally.

Axolotls Life Cycle

Axolotls have a unique and interesting life cycle. They start as eggs laid by the female and fertilized by the male. These eggs hatch into larvae after about two weeks. Unlike many amphibians, Axolotls remain aquatic and retain their gills throughout life. As larvae, they have external gills and a tail fin to help them swim. Over the next few months, they grow rapidly, developing legs and growing in size. Unlike other amphibians, axolotls do not undergo a complete metamorphosis and remain in their larval form even as adults, known as neoteny. They become sexually mature at around 6-12 months old but continue to look like large larvae with external gills and a fully aquatic lifestyle.

In inducing breeding conditions to encourage breeding, you need to simulate the conditions of their natural habitat. This involves lowering the water temperature to around 60°F (15°C) to mimic winter conditions, gradually raising it back to normal to simulate spring. This temperature shift can stimulate the breeding instinct. Additionally, ensure the tank is well-cycled and has clean water to support the health of both adults and their eggs.

Monitoring and Separating Pairs: Carefully monitor the pair during their time together to ensure that no aggressive behavior occurs. Once mating is successful and the eggs are fertilized, it's important to separate the female from the male to avoid potential aggression and to protect the eggs.

Setting Up a Breeding Tank

Choosing the Right Tank: A separate breeding tank is essential to provide a controlled environment for the axolotls during breeding. A 20–30-gallon tank is usually sufficient. The tank should be equipped with a sponge filter to provide gentle filtration without harming the delicate eggs.

Tank Setup

Substrate: Use a bare-bottom tank or fine sand, as gravel can be harmful if ingested.

Water Quality: Maintain pristine water conditions. Perform regular water changes and ensure the tank is well-cycled.

Temperature: Keep the water temperature consistent with your simulated breeding conditions.

Hiding Spots: Provide ample hiding spots for the female to feel secure and reduce stress.

Additional equipment: Ensure you have a thermometer to monitor water temperature accurately, a sponge filter for gentle filtration, and a water conditioner to remove any harmful chemicals from tap water.

Caring for Eggs and Larvae

Egg Care After successful mating, the female will lay eggs, which should be collected carefully and placed in a separate hatching tank. The eggs are typically laid in clusters and may take 10-14 days to hatch. Ensure the optimal water conditions and keep the eggs in a quiet, stable environment to reduce stress. Larvae Care Once hatched, the larvae, or baby Axolotls, are very small and require specific care

Feeding: Start by feeding them tiny live foods like infusoria or freshly hatched brine shrimp. Gradually introduce larger foods as they grow.

Water Quality: Maintain excellent water quality with frequent changes and gentle filtration.

Space: Provide ample space and hiding spots as they grow to prevent overcrowding and stress.

Monitoring Growth Regularly: Monitor the growth and development of the larvae. Ensure they are growing steadily and adjusting their diet as they become larger. Separate them by size if necessary to prevent cannibalism.

Raising Healthy Juveniles

Growth and Development: As the Axolotls grow into juveniles, continue to provide them with a clean and spacious environment. They should be transitioned to more appropriate juvenile food and given ample space to prevent overcrowding. Monitor their health and behavior regularly to catch any potential issues early. Socialization Juvenile: Axolotls can be housed together if they are of similar size to reduce aggression and stress. Provide plenty of hiding spots and monitor interactions to ensure they get along well.

Long-Term Care: Prepare for the transition from juvenile to adult care, ensuring that their tank environment and diet meet the needs of mature Axolotls. Continue with regular water testing and maintenance to keep their habitat healthy and conducive to their growth.

Chapter 8: Common Mistakes to Avoid and Frequently Asked Questions

Caring for Axolotls can be a rewarding experience, but it's essential to avoid common mistakes that can harm these delicate creatures. Here, we highlight the top 10 mistakes and how to avoid them, ensuring your axolotl thrives in its environment.

1. Inadequate Tank Size: Axolotls need plenty of space to move around comfortably. A common mistake is housing them in a tank that's too small. Always use a tank that's at least 20 gallons for one Axolotl, with an additional 10 gallons for each extra Axolotl.

2. Poor Water Quality: Maintaining clean water is crucial. Regularly test the water for ammonia, nitrite, and nitrate levels. Perform weekly water changes of about 20-30% and use a reliable water conditioner.

3. Incorrect Water Temperature: Axolotls flourish in cool water temperatures between 60-64°F (15-18°C). Avoid placing the tank in direct sunlight or near heat sources that can raise the temperature.

4. Overfeeding: Overfeeding can lead to obesity and water quality issues. Feed your Axolotl appropriately every 2-3 days, removing any uneaten food to prevent decay.

5. Using Gravel as Substrate: Gravel can be easily ingested by axolotls, leading to impaction and other health issues. Use fine sand or a bare bottom tank to avoid this risk.

6. Inadequate Filtration: Axolotls produce a significant amount of waste. Ensure your tank has a sponge or bubble filter to maintain water quality and provide adequate oxygenation.

7. Ignoring Health Signs: Axolotls can exhibit signs of stress or illness, such as loss of appetite, discoloration, or unusual swimming patterns. Regularly observe your axolotl and address any health concerns promptly.

8. Keeping Axolotls with Incompatible Tankmates: Axolotls should not be housed with fish or other animals that may nip at their gills or compete for food. They do best in a species-only tank.

9. Inadequate Hiding Spots: Axolotls need hiding spots to feel secure and reduce stress. Provide plenty of hides in their tank, such as caves or PVC pipes.

10. Neglecting Regular Maintenance

Frequently Asked Questions

1. How often should I feed my Axolotl? Young Axolotls should be fed daily, while adults can be fed every other day. Offer a varied diet of worms, pellets, and occasional treats like brine shrimp.

2. Can Axolotls live with other fish? Axolotls are best kept alone or with other Axolotls. Many fish species can nip at their gills or compete for food, causing stress and injury.

3. What should I do if my Axolotl is shedding its skin? Occasional shedding is normal. Ensure good water quality and monitor for signs of illness or infection if shedding is frequent or abnormal.

4. How can I tell if my Axolotl is stressed? Signs of stress include loss of appetite, frantic swimming, hiding excessively, and changes in color or gill condition. Check water quality and environmental conditions.

5. How do I transport my Axolotl safely? Use a clean, sturdy container with a secure lid filled with tank water. Avoid extreme temperatures during transport and handle the axolotl as little as possible.

6. Can Axolotls regenerate lost limbs? Yes, axolotls have remarkable regenerative abilities and can regrow lost limbs, gills, and even parts of their heart and brain.

7. What should I do if my Axolotl appears to be sick? Isolate the sick axolotl in a quarantine tank, check water quality, and consult a veterinarian experienced with amphibians. Early intervention can prevent serious health issues.

8. How long do Axolotls live? Axolotls can live up to 10-15 years in captivity with proper care. Maintaining good water quality and a suitable diet are key to their longevity.

9. Do Axolotls need a heater in their tank? Axolotls prefer cooler water temperatures, typically between 60-64°F (15-18°C). In most cases, a heater is not necessary unless the ambient temperature drops significantly.

10. Can I use tap water for my Axolotl tank? Yes, but you must use a water conditioner to remove chlorine and chloramine, which are harmful to Axolotls. Always treat tap water before adding it to the tank.

11. How can I distinguish between a male and female Axolotl? Mature males have longer, slimmer bodies and swollen cloacas compared to females. Females tend to have wider bodies and less pronounced cloacas.

12. What should I do if my Axolotl stops eating? Check the water quality and temperature first. Stress, illness, or changes in the environment can affect appetite. If the problem persists, consult a veterinarian.

13. Can I keep more than one axolotl in the same tank? Yes, axolotls can be kept together, but ensure the tank is large enough to provide ample space and hiding spots to prevent stress and aggression.

14. How do I clean my Axolotl tank? Perform regular partial water changes (10-20% weekly), remove uneaten food and waste daily, and clean the substrate, decorations, and filter media as needed.

15. What kind of substrate is best for Axolotls? Fine sand is ideal, as it is less likely to cause impaction if ingested. Avoid using gravel or small pebbles, which can be harmful if swallowed.

16. Can Axolotls climb out of their tank? Axolotls generally do not climb, but it is still a good idea to have a secure lid on the tank to prevent any escapes and to protect them from external dangers.

17. How do I acclimate a new Axolotl to its tank? Slowly introduce your Axolotl to the tank by floating its bag in the tank water for 15-20 minutes to equalize temperatures, then gradually add tank water to the bag before releasing the Axolotl.

18. Can Axolotls recognize their owners? While Axolotls do not have the same level of recognition as some other pets, they can become accustomed to their owner's presence and may respond to feeding routines and familiar movements.

19. What types of food are best for Axolotls? Axolotls enjoy a varied diet that includes live or frozen bloodworms, earthworms, brine shrimp, and high-quality Axolotl pellets. Avoid feeding them meat or fish meant for human consumption.

20. Why is my Axolotl turning pale? Paleness can be a sign of stress, illness, or poor water quality. Check the water parameters, ensure proper tank conditions, and observe your Axolotl for other signs of distress.

Regular tank maintenance is essential for a healthy Axolotl. Clean the substrate, decorations, and filter media as needed, and monitor water parameters regularly. Avoiding these common mistakes, you can ensure a healthy and happy environment for your Axolotl. Regular observation and maintenance will go a long way in providing the best care for these interesting creatures.

Chapter 9: Axolotls for Business

Starting Axolotls as a business venture can be both exciting and profitable. With their increasing popularity as pets and unique appeal, Axolotls offers a niche market that can be fulfilling if approached correctly. In this chapter, we'll study how to start, manage, and grow an Axolotl business, focusing on key aspects such as market research, breeding, sales, and customer care.

Market Research and Business Planning

Before jumping into the Axolotl business, conducting thorough market research is essential. Understanding the demand, competition, and target audience will help you create a solid business plan.

Identifying Your Market: Determine who your potential customers are. Are they hobbyists, educational institutions, or pet stores?

Assessing Competition: Research other Axolotl breeders and sellers in your area or online. What are their strengths and weaknesses?

Setting Goals and Objectives: Define clear, achievable goals for your business, such as the number of Axolotls to breed per month or revenue targets.

Setting Up a Breeding Facility

Creating a suitable environment for breeding Axolotls is crucial for producing healthy offspring and maintaining a successful business.

Tank Setup: Invest in high-quality tanks, filtration systems, and heating/cooling equipment to maintain optimal conditions for your Axolotls.

Breeding Pairs: Select healthy, genetically diverse breeding pairs to ensure robust offspring. Monitor their health and breeding behaviors closely.

Breeding Conditions: Simulate natural breeding conditions by adjusting water temperature and quality. Provide appropriate hiding spots for egg-laying.

Raising and Caring for Axolotl Larvae

Caring for Axolotl larvae requires attention to detail and consistency to ensure they grow into healthy juvenile

Egg Care: Once eggs are laid, transfer them to a separate, clean tank to prevent them from being eaten by adult Axolotls.

Feeding Larvae: Provide appropriate food for larvae, such as brine shrimp or daphnia, and gradually introduce larger food items as they grow.

Water Quality: Maintain pristine water conditions to prevent disease and promote healthy growth.

Marketing and Sales

Effective marketing strategies are essential to attract customers and grow your business.

Online Presence: Create a professional website and utilize social media platforms to showcase your Axolotls, share care tips, and connect with potential customers.

Photography and Videos: High-quality images and videos of your axolotls can significantly enhance your marketing efforts and attract more buyers.

Customer Engagement: Interact with your audience through Q&A sessions, informative posts, and regular updates about your breeding program.

Pricing and Sales Channels

Determining the right price and choosing the best sales channels are critical for business success.

Pricing Strategy: Set competitive prices based on your costs, market demand, and the quality of your Axolotls. Consider offering package deals or discounts for bulk purchases.

Sales Platforms: Utilize various sales channels, such as your website, online marketplaces, local pet stores, and aquarium clubs, to reach a broader audience.

Shipping and Delivery: Ensure you have a reliable and safe method for shipping axolotls to buyers, including proper packaging and fast delivery options.

Customer Care and Support

Providing excellent customer service is vital for building a loyal customer base and growing your reputation.

Pre-Sales Support: Offer detailed information and guidance to potential customers about axolotl care and setup requirements.

After-Sales Support: Provide ongoing customer support after purchase, including advice on feeding, tank maintenance, and health monitoring.

Feedback and Reviews: Encourage customers to leave reviews and feedback, which can help improve your services and attract new buyers.

Ethical Breeding and Business Practices

Maintaining ethical standards is essential for the well-being of your Axolotls and the integrity of your business.

Animal Welfare: Prioritize the health and welfare of your Axolotls by providing proper care, avoiding overbreeding, and ensuring humane treatment.

Transparency: Be transparent with your customers about your axolotls' origin, care, and health status.

Sustainability: To minimise your environmental impact, implement sustainable practices, such as recycling water and using eco-friendly materials.

Starting and managing an Axolotl business requires dedication, knowledge, and a passion for these unique creatures. By following best practices and maintaining high standards, you can create a successful and fulfilling venture that brings joy to axolotl enthusiasts and contributes to the conservation of this mysterious species.

Chapter 10: Fun Facts About Axolotls

Axolotls, also known as Mexican walking fish, are not just fascinating due to their unique appearance and regenerative abilities. Here are some fun and intriguing facts about these incredible creatures that will make you appreciate them even more.

The "Walking" Fish: Axolotls are often called "Mexican walking fish" because of their ability to walk on the bottom of their aquatic habitat using their four limbs. However, they are not fish but amphibians, specifically a type of salamander.

Eternal Youth: Axolotls are neotenic, meaning they retain their juvenile features throughout their lives. Unlike salamanders undergoing metamorphosis, Axolotls keep their gills and aquatic lifestyle, even as adults.

Incredible Regeneration: Axolotls are renowned for their regenerative abilities. They can regrow entire limbs, tails, and even parts of their heart, brain, and spinal cord. This remarkable trait makes them valuable subjects in scientific research.

Wide Range of Colors: Axolotls come in various color morphs, including wild type, leucistic, albino, golden albino, melanoid, and more. Each color morph has its unique beauty and characteristics.

Unique Breathing: Axolotls have both lungs and gills, allowing them to breathe in water and, to some extent, on land. Their feathery external gills are their primary breathing organs, but they can gulp air from the surface when necessary.

Sensitive to Light: Axolotls prefer dim environments and are sensitive to bright light. This is why providing shaded areas and hiding spots in their tank is crucial for their well-being.

Long Lifespan: With proper care, axolotls can live for 10-15 years or even longer in captivity. Their longevity makes them a long-term commitment for any pet owner.

Cultural Significance: Axolotls are native to Mexico, specifically the lake complex of Xochimilco. They hold cultural significance and are considered a symbol of Mexico's rich biodiversity. In Aztec mythology, the axolotl is associated with the god Xolotl, who transformed into an axolotl to avoid sacrifice.

Not Strong Swimmers: Axolotls are not strong swimmers despite living in water. They tend to "walk" along the bottom of their tank using their limbs rather than swimming like fish.

Axolotl Smiles: Axolotls are known for their "smiling" faces. Their upturned mouths give them a perpetual smile, making them endearing and popular among pet owners.

Axolotls are truly enthralling creatures with their unique biology and attractive behaviors. You might be a hobbyist, a researcher, or a curious pet owner, but these fun facts about Axolotls highlight just how special these animals are. Embracing the quirks and wonders of Axolotls can deepen your appreciation and commitment to their care and conservation.

Conclusion

Throughout this guide, we have studied the astonishing nature of Axolotls, from understanding their unique biology and setting up the ideal tank to providing proper care and even breeding these attractive creatures. Here are some of the key takeaways:

Understanding Axolotls: We learned about the history, natural habitat, and unique biological traits of Axolotls, including their remarkable regenerative abilities and their preference for dim environments.

Setting Up the Tank: We covered the essentials of creating a suitable home for your axolotl, including choosing the right tank, maintaining water quality, and selecting appropriate decorations and equipment.

Daily Care and Maintenance: Regular observation, water testing, and routine tank maintenance are crucial for keeping your Axolotl healthy and happy.

Feeding Your Axolotl: Understanding their dietary needs and providing a varied and balanced diet ensures your Axolotl's well-being.

Behavior and Interaction: Observing and understanding Axolotl behavior helps provide a stimulating environment and detect any health issues early.

Breeding Axolotls: For those interested in breeding, we covered the steps from setting up a breeding tank to caring for eggs and raising healthy juveniles.

Common Mistakes and Troubleshooting

Avoiding common pitfalls and knowing how to troubleshoot issues can prevent many problems and ensure a smooth Axolotl-keeping experience.

Axolotls for Business: We explained how to turn your Axolotl hobby into a business, including breeding, selling, and marketing.

Fun Facts and FAQs: Learning fun and interesting facts about Axolotls adds to the enjoyment of keeping these unique pets.

Encouragement and Final Tips

Keeping Axolotls can be a rewarding experience. These charming creatures bring a piece of the wild into your home and offer endless beauty with their behaviors and unique traits. Always provide a stable and clean environment, pay attention to their health and behavior, and enjoy caring for these extraordinary amphibians. For those new to Axolotl care, patience and diligence are key. It might seem overwhelming initially, but with time and practice, you'll become more confident in providing the best care for your Axolotls. Don't be afraid to seek advice from experienced keepers or join Axolotl communities online for support and tips.

Where to Find More Information

To continue expanding your knowledge about Axolotls, consider the following resources:

Online Communities: Join forums, social media groups, and online communities dedicated to Axolotl enthusiasts. These platforms are great for sharing experiences, asking questions, and finding support.

Scientific Journals: For those interested in the scientific aspects of Axolotls, research papers and journals can provide in-depth information on their biology, genetics, and regenerative capabilities.

Local Experts: Visit local aquariums, pet stores, or Axolotl breeders who can offer personalized advice and insights.

By continuing to educate yourself and staying engaged with the Axolotl-keeping community, you can ensure a fulfilling and successful journey with your Axolotls. Enjoy the journey, and happy Axolotl keeping!

www.ingramcontent.com/pod-product-compliance
Lightning Source LLC
Chambersburg PA
CBHW070352230526
45471CB00006B/2537